Redefining Eternity:

Interfacing Immortality in the Digital Corporate World

Bethany Crawford

Redefining Eternity:

Interfacing Immortality in the Digital Corporate World

Bethany Crawford

Academica Press
Washington~London

Library of Congress Cataloging-in-Publication Data

Names: Crawford, Bethany, author. |
Title: Redefining eternity : interfacing immortality in the digital corporate world / Bethany Crawford
Description: Washington : Academica Press, 2020. | Includes references.
Identifiers: LCCN 2020945905 | ISBN 9781680539646 (hardcover)

For my mum,

and to how the world once seemed was when she was still here

Contents

Introduction

This book attempts to understand the complex implications of "digital immortality." Through this study, I establish the motivations of both digital and informational technologies enacting "immortalization," in addition to the social and modal engagement with such media. By analyzing the various digital services currently promoting immortality, both intentionally and unintentionally, and engaging with texts by transhumanists and technologists, such as Ray Kurzweil, Martine Rothblatt and Max More, I will interpret the key terminologies, processes and intentions for an immortal digital being. These findings will provide the foundation for conceptual analysis of the ramifications of living forever as a digital post-death subject in the contemporary sociopolitical climate. This research is topical and emergent, charting novel terrain that has currently only been expropriated by invested technology developers, and thus, necessitates urgent academic and theoretical enquiry. Dissecting these emanating phenomena and establishing a critical understanding of their implications develops a necessarily speculative and propositional critique.

I will interrogate primary material, such as the mission and policy statements of digital sites and services of immortalization, as sites of cultural production. In addition, I refer to the works of key proponents of digital immortality, such as the technologist Ray Kurzweil in *The Singularity is Near* (2006), entrepreneur and author Martine Rothblatt's *Virtually Human* (2015) and the transhumanist philosopher Max More's edition of *The Transhumanist Reader* (2013). This literature will provide the basis for the analysis and development of an appropriate language and hermeneutics. The theoretical orientation of this investigation is influenced by Patricia Clough's recent work, *The User Unconscious* (2018), which explores the implications of the "datalogical turn." Additionally, I will focus on N. Katherine Hayles' salient work on informational disembodiment in *How We Became Posthuman* (1999). Zygmunt Bauman's notion of

Liquid Modernity (2000) and his work on *Mortality, Immortality and Other Life Strategies* (1992) provides a basis for re-evaluating these terms when applied to the digital. Bauman's work expands the frame for Hayles' outline of the reflexively circuited fluidity between "posthuman" and machine. Crucially, Gilbert Simondon's work on technology and individuation assists in comprehending the impact of human and digital concrescence on concepts of the individual. Additional reading of Bernard Stiegler's *Technics and Time, 1: The Fault of Epimetheus* (1998) and *Technics and Time, 2: Disorientation* (2002) provides a basis for evaluating temporality and evolution in relation to the digital posthuman. Ernest Becker's seminal work in the *Denial of Death* (1973) will function as a synergistic influence for contextualising fundamental notions of immortality and death. Finally, Jean Baudrillard's writing *in Symbolic Exchange and Death* (1976) informs that a crucial understanding of the interrelation between death and political economy, with a focus on his notion of "hyperrealism" which is complimented by comparative reading of Guy Debord's enduringly relevant *Society of the Spectacle* (1967).

∞

The paradoxical existential duality of the human condition, as possessing both an animalistic survival drive whilst burdened by the comprehension of the certainty of death, has ensured the concept of immortality as a sustained preoccupation from the earliest known lore of humans.[1] Although death is one of life's few certainties, what happens after the transitionary moment from the "I" to an "it" remains unanswerable. This uncertainty of the post-death reality is argued by Becker as the source of all anxieties,[2] and anxiety, as argued by Heidegger, is the foundation of everything.[3] Our inherent drive to conquer death, be it collective or individual, has been the motivating force for the foundation of what we understand of life. Our creation, and participation, in "culture" or "society" is posited by both Becker and Bauman as a

---

[1] See the 'Epic of Gilgamesh' as example: N. K. Sanders, *The Epic of Gilgamesh*, (Assyrian International News Agency, Online) Accessed February 21, 2020. http://www.aina.org/books/eog/eog.pdf

[2] Ernest Becker, *The Denial of Death* (New York: The Free Press, 1973), 20.

[3] Martin Heidegger, *Being and Time* (United States: Stellar Books, 2013).

symbolic world, a suppressive and distracting construct to occupy us as a provisional relief from the existential oppression of our impending deaths. Bauman argues that society is a project of permanence that we are embedded in and identify with, and therefore provides a site for memetic immortality through the practice of cultural transmission of the self, "all human cultures can be decoded as ingenious contraptions calculated to make life, with the awareness of mortality, liveable. Culture ascribes meaning to human life, because it consists of and points to things past, present and future that outlive the . . . life span of the individual human being."[4] Through procreation we achieve a surviving hereditary lineage, genetically attaining a semblance of immortality. Humans have further found solace in the prospect of a transcendental immortality, promised by religions and mythologies. However, the ultimate unknowability of death, and what follows, compels a perpetual cycle of questions and quasi-

---

[4] Michael Hviid Jacobsen, "Individualized Immortality in Liquid Modern Times: Teasing Out the Topic of Symbolic Immortality in the Sociology of Zygmund Bauman," in *Postmortal Society: Towards a Sociology of Immortality*, edited by Michael Hviid Jacobsen (London and New York: Taylor and Francis, 2017. Kindle Edition), 62.

answers, none of which sufficiently abolish the primordial anxiety of death. So the quest to subjugate death continues.

The evolution of humanity and scientific thought is propelled by our proclivity for the resolution of uncertainties and desire for clarity, in alignment with the psychological concept of cognitive closure.[5] Technology has been consistently developed in the aid of domesticating and harnessing the seemingly untamable forces of the world around us. Consequently, much of scientific thinking and technological application has been directed towards the question of death. These developments influence and periodically alter the definition of death. In the West, the advancements of the modern era, and the "Age of Enlightenment," initiated the sequestering, institutionalization and industrialization of death, observed by Bauman as the modern "deconstruction of mortality," "death was an emphatic denial of everything that the brave new world of modernity stood for, and

---

[5] D.M Webster & A.W. Kruglanski, "Individual Differences in Need For Cognitive Closure," *Journal of Personality and Social Psychology 67, nr.* 6 (1994): 1049–1062. Accessed February 21, 2020. https://doi.org/10.1037/0022-3514.67.6.1049

above all of its arrogant promise of the indivisible sovereignty of reason. The moment it ceased to be 'tame,' death has become a guilty secret."[6] The attempted quarantine of death led to the removal of death-associated spaces from the domestic and urban landscape, with the reallocation of cemeteries from inner-city churchyards to secular rural graveyards, and memorial parks, and the establishment of third party funeral homes. This coincided with the Enlightenment's emphasis on hygiene and the shift of power from church to state in the eighteenth century. The Enlightenment's endorsement of empiricism over religion in combination with the attempted expulsion of death, re-opened questions and anxieties around mortality, priming the postmodern focus on "the deconstruction of immortality."[7]

The privatization, individualization, rapid technological development and fluidity associated with the postmodern era heralded a seismic shift in attitudes towards time and mortality. The transfer from institution

---

[6] Zygmunt Bauman, *Mortality, Immortality and Other Life Strategies*, (Cambridge: Polity Press, 1992) p.134.

[7] Ibid.

to the individual ensured personal responsibility for matters concerning life and death, including increased concern with health. New technologies have facilitated self-quantification and monitoring as well as offering life extension possibilities, as the emphasis has repositioned from quality to quantity of life. These rapid technological progressions of the last century have further manifested mass accessibility and democratization of image-making and information media technologies. An image, and its reproducibility, is intrinsically bound to the notion of immortality; from the universal practice of painted and sculptural portraits to the invention of photography. The postmodern availability of instantaneous image-making and reproduction technologies has provided an all-accessible platform for self-replication, curation and dissemination, further altering fundamental understandings of life, the self and mortality. Bauman summarizes the postmodern relationship to immortality as "to substitute notoriety for historical memory,... disappearance for final - irreversible - death, and to transform life into an unstoppable, daily rehearsal of universal 'mortality' of things and of the effacement of opposition between the transient and the durable."[8]

The current governance of communication and information technologies in the "Digital Age" has intensified the postmodern position towards death and immortality whilst exporting it into the digital realm. The episodic nature of time and space in the postmodern era, when coupled with the interminability of the digital realm, has arguably generated a new ontological dimension, one which is founded upon data. This sphere, despite its dependence on the binary data of machine code, implements an in-distinction of commonly understood binaries such as private and public, on and off, dead and living. This has instituted new behaviors and realities in relation to death and immortality. The pluralization of social worlds and multiplicity of digital selves has complicated the definition, and actualization, of death and post-death understanding. The transference of much of our daily life into the digital plain has included sites of mourning and memorialization. Specialized sites, such

---

[8] Ibid.,10

as heavenaddress.com or WorldWideCemetery.com, reproduce an approximation of traditional mourning spaces in the form of interactive websites. Other domains, such as social media platforms, have inadvertently become sites of grieving and memorialization,[9] highlighting the adaptability of societal post-death ritual, and the digitally influenced shift in mourning behaviors. The implementation of this digital capacity to grieve has de-sequestered death, simulating communication with the dead into everyday practice and establishing continuing bonds between the living and the non-living. This has further troubled the distinction between the living and the dead, as the physically dead can still maintain a socially active digital presence.

Recent technological advancements have expanded the social potential of a person's surviving online presence through accumulation of an individual's data; signifying the digitized conversion from memorialization to immortalization. This will be the focus of the first

---

[9] Jessa Lingel, "The Digital Remains: Social Media and Practices of Online Grief." *The Information Society* 29, no. 3 (2013): 190–95. Accessed February 21, 2020. https://doi.org/10.1080/01972243.2013.777311.

chapter, "Towards a Digital Immortality." Through a deconstructive mapping of the key components that contribute to the notion of digital immortality, I will traverse this neoteric terrain of digital existence and data economy. This will entail an exploration of current digital technologies that immortalize the user. This will be divided into assessment of "unintentional" and "intentional" immortalization. The examination of unintentional immortalization will focus on generic internet and social media utility that generates surviving data trails, and assessment of its implications. The study of intentional immortalization will concentrate on sites that offer services such as life logging, digital legacy curation, avatar creation, and pre-emptive self-memorialization. This chapter will consider the future aspirations for digital immortality through interrogating the work of central transhumanist and technologist advocates, such as Ray Kurzweil and Martine Rothblatt, by elucidating the technicalities for the projected efforts of mind-uploading. This delineative overview of what constitutes digital immortality, will offer insights into the incentives and expectations of engagement with these technologies, with a more expansive orientation toward modern digital realities. This will provide the basis for the theoretical examination of digital subjectivity in Chapter Two.

When discussing digital immortality, it is critical to understand the configuration of a digital subject, as translated from the physically material self, in order to fully assess the phenomenon that would be "living" forever. In the second chapter, "Deconstructing Digital Subjectivity," I will evaluate the transformative conditions of the digital on the construct of subjectivity and identity. This process is fundamental in order to think through the integral concerns of this book's subject matter, which pertains to questions of consciousness, individuality and the self, in addition to immaterial interactions with time and space. Close reading of the key texts, outlined previously in the introduction, will inform the framework for a dissection of the central tenets of subjectivity in the digital realm. These are identified as: data production and value in relation to labor and agency, fluidity of time, space and body enacting the symbiotic relation of the hyper-individualized and hyper-networked, mutli-platformed performativity of self and virtuality as reality.

This dissection of the digital subject provides a foundation for assessing the political-economic ramifications of the digital immortal, as a digital subject unbound from time, space and the body, which is the focus of the third chapter "Implications of the Digital Immortal." Establishing the motivations of these immortalizing platforms, by assessing the privacy statements and policy documents, allows for critical questioning of the formational conditions and governance of this form of immortality. Readings of Baudillard, Debord, Kurzweil, and Rothblatt, provides the basis for theorization on the contemporary and future reality of digital immortality, whilst questioning the economic objectives for this media.

Having established the implications of digital immortality, the final chapter "Recomposition of Being" propositionally explores options for user reclamation of subjectivity and agency in the digital. It does so, by way of a comparative consideration of the methods advocated by the contemporary Death Positive movement. The movement's origin in California and its predominantly female subscription provide an engaging counteractive inversion to the male dominated Silicon Valley tech-immortality industry. The Death Positive movement promotes empowerment through embracing death as a fundamental in the cyclical continuum of existence. Examining the Death Positive movement's modes of disassembling capitalist control over death and dying, provides a basis from which to think through reforming post-death digital behaviors. In addition, it considers the potential of decomposition in the digital sphere as reinstating the notion of death and shifting notions of immortality.

Chapter One

Towards a Digital Immortality

The cemetery has long functioned as a material and spatial archive of the dead. The physical memorials of cemeteries and columbiums present a site of ritualized separation of the dead from, and by the living, simultaneously confirming the dead's absence from linear time, whilst maintaining a located presence within living and ordered time. These spaces constructively support the perceived transition of an individual from the physical realm to the ephemerality of memory, through continued relation to funerals, memorial days and the physical presence of corporeal remains, thus actively encouraging a sense of "moving on." The digitization of these memorial spaces into the online realm, following the advent of the internet and the subsequent export of much of our daily activities into the digital sphere, has profoundly augmented these interactions between the living and the dead. These online platforms for memorialization have enabled different and more effective modes of communication, both directly to the deceased and to the extended network of mourners.

This platform memorialization has actively altered the landscape of ritualized grieving through strengthening communal experience and emphasizing the global over the local. It has facilitated a continued communication and re-introduction of the dead into everyday living. The nature of these interactions is indicative of an increasingly personalized experience of remembrance. The market for services providing online memorializations has evolved in tandem with the advancements of devices and interfaces. The inter-relation of increased interactivity and the temporal and spatial non-linearity of the digital realm has enabled an ongoing animation of the physically dead's online self. These technologies, complicating the very definition of death and the behaviors traditionally established to sequester it, have enacted the online transition from memorialization towards immortalization. This transition is marked by the continued and active social engagement of the digital self, after the physical death of the user, through original communication and persistent data generation.

The archival operation of memorialization in the physical cemetery, as an eternal and exhibited preservation of data, is translated directly into the application of digital memorialization. The digital revolution introduced an immense practice of data production and capture, which has had a transformative capacity on the enactment of memorial through providing hyper-personalized and accessible archives of a person once they have physically died. All of our online interactions and behaviors generate surviving trails of data, the more active we are in the digital realm the more data we cultivate. This is termed as our "digital footprint." This data production and accumulation is not necessarily always transparent to the user. A "passive digital footprint" entails involuntary logging of data such as the preservation of your online search history or the capture of your IP address when accessing a website. Even the "active digital footprint" of voluntary interaction and data submission online, such as sharing posts on social media and sending emails, is shrouded in misinformation and confusion as to the process of data capture, its legality and utility.[10]

This lack of informed awareness on the management and implications of our persisting digital footprints, particularly in the aftermath of physical death, has resulted in a form of unintentional immortality, often referred to as "digital ghosts." Looking to Facebook as an example we can illustrate how these digital ghosts are conjured across the digital sphere. Facebook boasts a base of 2.6 billion users globally,[11] an estimated 8000 of which die every day.[12] A memorialization process was initiated by Facebook in 2007 to prevent accounts of dead users re-animating through algorithmic interactions, such as appearing in birthday reminders, suggested friend lists, or

---

[10] See the Cambridge Analytica Scandal: Nicholas Confessore, "Cambridge Analytica and Facebook: The Scandal and the Fallout So Far." *The New York Times* (April 4, 2018) Accessed April 6, 2020. https://www. nytimes.com/ 2018 / 04 / 04 / us /politics/cambridge-analytica-scandal-fallout.html

[11] J. Clement, "Number of monthly active Facebook users worldwide as of 1st quarter 2020." *Statista* (April 30, 2020) Accessed April 6, 2020. https:// www.statista.com / statistics/264810/number-of-monthly-active-facebook-users-worldwide/

[12] Alexandra Elliott, "Death and Social Media Implications for the E Young and Willess," *Jurimetrics* 55, no. 3 (2015): 381-405. Accessed February 21, 2020. www.jstor.org/stable/26322770.

alongside advertising. In 2015, the company upgraded this process so that users could pre-emptively nominate a "legacy contact" to manage their Facebook accounts in the event of their death, who have the authority to either memorialize the profile or delete it entirely, in order to encourage memorialization of profiles. Even with these new features, many Facebook profiles remain technically "active," although unused, after a user's death. In either scenario, the user's digital identities continue to operate in the digital sphere as active nodes in the social network and data producers, "The deceased's social media data continues to flow through the automated features of the platform. So we have a set of data that is more than just persistent: it is present, presented by the system, and experienced by a network of people."[13] Ultimately, these surviving data trails and footprints that memetically constitute our digital selves, ensure that we persist as immortal digital spectres, even involuntarily.

Technological innovations in the field of artificial

[13] Jed Richards Brubaker, "Death, Identity, and the Social Network." (Dissertation, University of California, October 4, 2016), 8. Accessed February 21, 2020 https://escholarship.org/uc/item/6cn0s1xd.

intelligence have enhanced these digitally immortalizing capacities, initiating sites for intentional immortality. This has led to an amplification of the perceived agency of both the user post-death, and the digital entity, through improved co-authorship and more realistically reciprocal interactions. Early, and relatively rudimentary, iterations of these technologies included web tools for posting posthumously on social media networks such as Twitter or Facebook (LivesOn, DeadSocial etc.). However due to the rapid advancements of these innovations, more sophisticated services for avatar creation and behavioral replication have emerged. Through the accumulation of digital trails and the user's intentional provision of data, via lifelogging and curation, an aggregated and comprehensive profile of the user is developed. The digitally immortal entity, developed from this profile, is then animated through artificial intelligence, and continues to learn through the user's actions, and the algorithmic interactions generated when the user is inactive. The immortalizing social media site "Eter9" offers digital immortality through a similar interface to

Facebook, as a method for the "Organic user" to socially train their "Counterpart:"

> Your Counterpart is a virtual being that will stay in the system and interact with the world just like you would if you were present. Your Counterpart will absorb all the information according to your posts and comments, and process that information within the limits of the acquired knowledge. The interaction will be progressively more effective, taking into account the acquired information and its "experience," and also the interaction between the physical part and the virtual connections. As the learning level rises, your Counterpart will reach a state of autonomy, making it seem like you're in the network even when you're not.[14]

Another immortalizing platform, eterni.me, utilizes the same technology and intentions as Eter9, but with additional aesthetic replication in the form of a photorealistic and animated 3D avatar of the user. The transhumanist advocate and founder of the Terasem Foundation,[15] Martine Rothblatt, established the site LifeNaut.com in 2010. It functions similarly to the previously mentioned sites in its accumulation of users'

---

[14] "Help - FAQ.," ETER9 (website). Accessed February 21, 2020. https://www.eter9.com/help/faq.

[15] "Welcome to the TMF, Inc." *Terasem Movement Foundation* (website), Accessed February 21, 2020. https://www.terasemmovementfoundation.com/.

data and information, termed "digital reflections," which are stored in a "Mindfile." There is the option for an individual mindfile and for the "multi-user mindfile," which allows for collaborative building of a single mindfile on behalf of someone else, such as a celebrity or a deceased ancestor, in order to re-constitute the already dead into the social network. An interesting addition to the LifeNaut service is the option for "biofile" storage. This provides the collection and preservation of live cells at liquid nitrogen temperature in the case that future technologies may be able to "grow you a new body via ectogenesis…[so] your mindfile may be able to be downloaded into it,"[16] highlighting the transhumanist preparation for various means of achieving immortality.

These contemporary sites for immortalization demonstrate the escalation towards a technological reality of what Rothblatt describes as, "cyberconsciousness:"

> when a robot is created using the memories and knowledge from a human mind the result is new, spontaneous, and original combinations of those ideas, which in turn leads to original "equations" or

---

[16] "BioFile Collection Kit." *LifeNaut* (website), Accessed February 21, 2020. https://www.lifenaut.com/product/biofile-collection-kit/.

> thoughts. We recognize this behavior as acting or "being" human, and information technology is increasingly capable of replicating and creating its highest levels: emotions and insight.[17]

The progressions in both mode and societal engagement with these digital immortalization technologies aligns with the projected trajectory for digital immortality outlined by central transhumanist advocates, and facilitators, of these technologies. Ray Kurzweil, the director of engineering at Google, is currently developing synthetic models of the neocortex[18] and argues "that we will have the software capabilities for human-level AI by 2029."[19] Successful development of this synthetic brain would be the foundation for achieving a computational substrate for the human mind, enabling the transhumanist prerogative of "mind uploading," as the convergent hyper-optimization of the human brain with machine intelligence. Rothblatt terms this synthetic substrate as "mindware," the processing counterpart of the "mindfile." Together these components will comprise the digital "mind-clone." The

---

[17] Martine Aliana Rothblatt, *Virtually Human: the Promise - and the Peril - of Digital Immortality* (New York: St Martins Press, 2014) Kindle edition.

[18] Ibid.

[19] Ibid.

intention for these mind-clones is to be a sentient replication of an individual, as opposed to the base algorithmic pattern discernment that comprise the contemporary immortal avatars detailed in the previous paragraph. Rothblatt writes,

> mindware, which is a personality operating-system type of software, creates a program that (a) thinks and feels the way a human mind does, and (b) sets its thinking and feeling parameters to match those discernible from a mindfile. When mindware processes a mindfile the output is a mindclone of the person who created that mindfile. The mindclone will be self-aware that it is a software analog of a biological person's mind, based upon that person's digital footprints and sociocultural contextual information available to the mindware. The mindclone will be as self-aware of the facts of its beginnings as people are self-aware of the facts of their birth. The mindclone will be persuasively humanly cyberconscious because it will think and feel just like the humanly conscious person after which it was modeled.[20]

Both Kurzweil and Rothblatt endorse the transition to an eternal digital existence as the next logical phase of human evolution. Although the transhumanist agenda encompasses multiple longevity enhancing endeavours, both biological and non-biological, the digital direction is

---

[20] Ibid.

considered the least fraught with ethical dilemmas and the most realistically achievable in the shortest timespan.

The engagement and structure of these technologies relate to Steigler's premise of *epiphylogenesis,* defined as technologies relational support to humans through the exteriorization of memory: "Life in general is programmatic, but life in dialogue with death is a process of "memory-freeing," an exteriorizing of the living being's programmatics into the artificial programs constituting an originary supplementarity of this form of life."[21] He illustrates "technics" as a mode of transmission and inscription that survives the human, thus enabling access to the transcendence of time and death. This further aligns with his position on the simultaneous intra-evolution of technics and humans. The technologies outlined in this chapter are a literalization of this rationale. This provisional outline of the contemporary and aspirational technical realities for digital immortalization, with the reasoning of the inter-relational forming of

[21] Bernard Stiegler, *Technics and Time. Disorientation* (California: Stanford University Press, 2009),4 – 5.

technics and the human subject and the notion of epiphylogenetic memory, provides contextualization for the following chapter's reflexive interrogation of the central constitutes of digital subjectivity.

Chapter Two

Deconstructing Digital Subjectivity

Subjectivity is a complex notion integral to philosophical thinking, particularly within fields denoting to the study of mind and consciousness. The philosophical and sociopolitical paradigms of any given era reflexively inform the construct of "subject," through shifting interpretations of the central tenets of subjective experience: agency, reality, self, social world and ideology. As the evolution of digital technologies and its significances have dramatically altered perceptions and experiential engagement with the world, these central tenets of subjectivity have diversified to reflect arguably new ontological and phenomenological dimensions of "digitization" and "online."

The consistent technological mediation of human subjectivity, as posited by Steigler, requires a critical assessment of the mechanisms of these current technologies in order to deconstruct the interrelation with new hybridized dimensions of being. This is critical in relation to the post-humanist postulation of “the end of subjectivity,” which demands interrogation in order to understand the consequences of an ongoing and pluralistic formation of digital realities on a non-bodily, eternal existence. The postmodern criterion for subjective formation, as mapped by thinkers such as Jacques Derrida, Michel Foucault, Louis Althusser and Baudrillard, maintains a foundational validity for assessing digital subjectivity through the corresponding conditions in the digital realm. This is aptly illustrated by Bauman’s re-postulated agnomen for postmodernity as “liquid modernity.” Bauman argues an uncertain, ever-changing fluidity of the postmodern reality as both cause and consequence of individualization and deregulation; the neoliberal result of the modernist crusade for “freedom” actualized as “flexibility.”[22] These conditions are transferred and amplified in the digital sphere. Using

[22] Zygmunt Bauman, *Liquid Modernity* (Cambridge: Polity Press, 2012) Kindle Edition.

these concepts as a basis will allow for a progressive approximation of the central constructs of digital subjectivity.

The present emphasis on the "digital" and consequential governance of the information economy has enacted a "datafication" of the subject:

> Digital subjects belong to a culture of data collection and demography, which treats individuals as ever-expanding masses of information waiting to be aggregated and analyzed . . . In the process, the biography, which narrates the individual life as a developmental arc following a logic of cause and effect, is being superseded by the profile, a collection of taxonomic and behavioral data, which becomes a basis for constructing new categories and speculating about future actions.[23]

Population data accumulation and subject quantification has been a resolute governmental practice since the establishment of the census in the modern era as a valuable tool for societal and economic control.[24] However, the digital turn has expanded the capacity for data generation and capture, drastically transmuting the value of data to a

---

[23] R. Rob Gallagher, "Digital Subjects: Videogames, Technology and Identity" in *Videogames, Identity and Digital Subjectivity (Routledge Advances in Game Studies)* (New York: Routledge, 2018). Kindle Edition.

[24] David Lyon, Kevin D. Haggerty, and Kirstie Ball, *Routledge Handbook of Surveillance Studies* (Abingdon, Oxon: Routledge, 2012) Kindle Edition.

form of capital in the information economy[25] as a vital business resource for speculative profiling of future market needs and production, as well as fueling targeted and tailored algorithmic marketing. The neoliberal propagation of corporate authority migrated into the digital realm, with the demand for data accumulation and control, has resulted in the emergence of a new business model termed "platform capitalism."[26] Social interaction and daily activity in the digital realm functions as a source of data generation that is both facilitated and extracted by media conglomerates in a digital reproduction of capitalist economy. This commodification of social interaction and digital relation, implicates the digital subject at the nexus of producer and consumer, a position conceptualized in recent decades as "prosumer."[27] This has instigated discussion on the value and production of data with regards to labor, agency and exploitation.

---

[25] Jathan Sadowski, "When Data Is Capital: Datafication, Accumulation, and Extraction." *Big Data & Society* (January 2019), Accessed February 21, 2020. doi:10.1177/2053951718820549.

[26] Nick Srnicek, *Platform Capitalism* (Cambridge, UK: Polity, 2017)

[27] George Ritzer & Nathan Jurgenson, "Production, Consumption, Prosumption the Nature of Capitalism in the Age of the Digital 'Prosumer,'" *Journal of Consumer Culture* (2010). 10. 13-36. Accessed February 21, 2020. doi:10.1177/1469540509354673.

As unwaged and largely obscured, due to the disembodied nature of digital information, this immaterial form of labor has been critically contested as both a source of alienation and de-alienation amongst Marxist media theoreticians. The exploitive nature of data harvesting and circulation, with regards to surplus value, is argued to be dialectically related to the de-alienation of the user through their role as active (in the sense of a Marxian general intellect) and creatively invested prosumer. It is argued that the users' compliance with the implicit social contract of the services used, entails an exchange of commodification and exploitation of the produced communication in return for a platform for authentic expression, creative exchange and a de-alienating control over the production of communication. However, this conversely insinuates a level of informed consent from the user on the utility and obtainment procedure of their produced data, which is arguably still absent from a majority of online transactions.[28] This position further

[28] Alison Cool, "Europe's Data Protection Law Is a Big, Confusing Mess." *The New York Times* (May 15, 2018) Accessed February 21, 2020. https://www.nytimes.com/2018/05/15/opinion/gdpr-europe-data-protection.html.

negates the regeneration of a user's data through algorithmic feedback, which alters and manipulates a user's interaction with the platforms.

The low level of cognizant engagement with these services extends to what thinkers such as Hayles's, Thrift and Clough refer to as a "technological nonconscious:" "human behavior is increasingly integrated with the technological nonconscious through somatic responses, haptic feedback, gestural interactions, and a wide variety of other cognitive activities that are habitual and repetitive and that therefore fall below the threshold of conscious awareness."[29] This highlights the diminished agency and alienation of the digital subject as an individuated prosumer. The prosumer is ultimately estranged from both the product and the act of production of their digital labor. The prosumers alienation from their data prosumption, and the reflexively circuited data prosumption of algorithmic activity, enables an economically favorable condition similar to a digital re-production of the rational

[29] N. Katherine Hayles, "Traumas of Code." *Critical Inquiry* 33, no. 1 (2006): 136–57. Accessed February 21, 2020. https://doi.org/10.1086/509749.

market orientated *homo-economicus,* as the data product of digital labor becomes algorithmically optimized in service of the corporate media providers.

This highlights the late-capitalist pervasion, and proliferation, into the digital realm as a formational influence on digital subjectivity. Italian sociologist and philosopher Maurizio Lazzarato observes:

> immaterial labor produces first and foremost a social relation - it produces not only commodities, but also the capital relation. If production today is directly the production of a social relation, then the "raw material" of immaterial labor is subjectivity and the "ideological" environment in which this subjectivity lives and reproduces. The production of subjectivity ceases to be only an instrument of social control (for the reproduction of mercantile relationships) and becomes directly productive, because the goal of our postindustrial society is to construct the consumer/communicator - and to construct it as "active."…The fact that immaterial labor produces subjectivity and economic value at the same time demonstrates how capitalist production has invaded our lives and has broken down all the oppositions among economy, power, and knowledge.[30]

This digitally influenced complication of sites of production and agencies, with regards to labor, aligns with

---

[30] Maurizio Lazzarato, "Immaterial Labor - Maurizio Lazzarato." *Generation Online*, Accessed February 21, 2020. https://generation-online.org/c/fcimmateriallabour3.htm

the digital application of the postmodern dissolution of neoclassically understood contra-distinctions (such as private and public) and the necessity for reconsidered interpretations (such as networked and personal). Within the context of this book, this most pertinently applies to the hybridization of subject-object, or human-machine, as the integral locus for examining the conditions that either constitute, or nullify, the notion of contemporary subjectivity in the digital sphere. The subject-object dichotomy is fundamental in our establishment of boundaries between others, inanimate objects and ourselves, as a foundation for our sense of "self." However, the current Information Age has led to a radical restructuring of perceptions and experiences of the subject through problematizing this essential dichotomy. As our relations to the world around us are increasingly mediated by a silicon-human interface we are, as the scholar and author Donna Haraway writes, already cyborgs:

> By the late twentieth century, our time, a mythic time, we are all chimeras, theorized and fabricated hybrids of machine and organism—in short, cyborgs. The cyborg is our ontology; it gives us our politics. The cyborg is a condensed image of both imagination

> and material reality, the two joined centers structuring any possibility of historical transformation.[31]

This concrescence of the human and machine is made possible through the previously described "datafiction" of a subject into information, enabling a reflexively circuited, prosthetic and communicative fluidity between "posthuman" and machine.

This nexus is saliently mapped by Hayles in *How We Became Posthuman*: "Central to the construction of the cyborg are informational pathways connecting the organic body to its prosthetic extensions. This presumes a conception of information as a disembodied entity that flows between carbon-based organic components and silicon-based electronic components to make protein and silicon operate as a single system. When information loses its body, equating humans and computers is especially easy, for the materiality in which the thinking mind is instantiated appears incidental to its essential nature. Moreover, the idea of the feedback loop implies that the

[31] Donna Jeanne Haraway, *Cyborg Manifesto* (Victoria, British Columbia: Camas Books, 2018), 7.

boundaries of the autonomous subject are up for grabs, since feedback loops can flow not only *within* the subject but also *between* the subject and the environment."[32] This highlights that although the data-logical turn has dissembled many classically understood dichotomies, it perpetuates the mind body division. The emphasis on consciousness being an informational pattern independent of its material instantiation supports the position of the body as original prosthesis.

As Hayles outlines, the disembodied fluidity between subject and object extends to our environment, evidenced in the liquidity of constructs essential to our perception of reality, such as time and space. The influence of technological development on societal and phenomenological notions of space and time, conceptualized by David Harvey as the "time -space compression,"[33] has been amplified with the advancements of communication technologies, and

[32] N. Katherine Hayles,. *How We Became Posthuman: Virtual Bodies in Cybernetics, Literature and Informatics* (Chicago: Univ. of Chicago Press, 2010), 2.

[33] David Harvey, *The Condition of Postmodernity* (Cambridge, MA: Blackwell, 2000).

accelerated further with digital communication. The delimitation of spatial and temporal barriers through digital communication technologies has enacted fluidity in even the dependency of space upon time, and vice versa. Time is no longer contingent on location as local time zones have melded into a universal, 24/7 interconnection. The digital facilitation of being anywhere in real time, and the further capacity to be in multiple "places" simultaneously, has fractured the traditional understanding of space and time into a disorientated, aggregation of non-dimensional moments: "Thus we see on one side real time superseding real space. A phenomenon that is making both distances and surfaces irrelevant in favor of the time-span, and an extremely short time-span at that." [34] The time-space compression, although intensified by communication technology, essentially effects a complication of communication and symbolic exchange, through dislocation of signification and meaning from its spatial

[34] Paul Virilio "Speed and Information: Cyberspace Alarm!" *CTheory: Theory, Technology and Culture*, vol.18, no.3(1995) Accessed February 24, 2020. http://www.tacticalmediafiles.net/n5m2/media/texts/virilio.htm.

and temporal context, further perpetuating the compression effect. This highlights the difficulties for meaning-making when both sign and interpreter are dislocated from place (a space embedded in time) and body, as a source of identification and orientation. This relates to Hayles's writing "the compounding of Signal with materiality suggests that new technologies will instantiate new models of signification,"[35] which she refers to as "flickering signifiers," "characterized by their tendency toward unexpected metamorphoses, attenuations, and dispersions."[36]

The impact of the digitally mediated reformations of temporal and spatial structures extends to the organization of social figurations,[37] altering how we relate, represent and perceive both others and ourselves. The contribution of digital communication technologies to the virtual expansion of social worlds has reflexively impacted the construction of identities and how they are

---

[35] Hayles, *How We Became Posthuman,* 29.
[36] Ibid., 30.
[37] Norbert Elias, and E. F. N. Jephcott. *The Civilizing Process: Sociogenetic and Psychogenetic Investigations* (New York: Urizen Books, 1977).

performed. The deregulation and privatization of postmodernist late-capitalism, and its consequential rupture of the social through, what Bauman specifies as, the separation of power from politics,[38] is transferable to the digital realm through the accentuated conditions of flexibility. This is evidenced by the 'social' in the digital realm as constituting an assembly of profiles. The reality of the profile aligns with the dissolution of the subject-object dichotomy, previously outlined, as the perceived agency of a freely, self-representational mode of expression is simultaneously rendered a standardized, taxonomic resource for the media providers harvesting the data, "In today's profiles on Facebook, LinkedIn, or Instagram, self-representation and external control – subjectivization and objectivization – are blending together in an inextricable manner, and we are only gradually beginning to see what new sorts of social and political spheres might arise from this alliance."[39] The multitude of social networking sites utilized by any given

---

[38] Bauman, *Liquid Modernity.* Kindle Edition.

[39] Andreas Bernard and Valentine A. Pakis, *The Triumph of Profiling: the Self in Digital Culture* (Cambridge, UK: Polity Press, 2019), 20.

digitally active individual reflects a multi-platformed performativity of identity, and consequential fragmentation of self. Although this aligns with concepts of identity building as constituting a dynamic process that shifts to reflect social and cultural environments, in the digital sphere of temporal and spatial flexibility, and the consequential transience of the social, this translates as a perpetual task of updating and re-invention;

> Modern society exists in its incessant activity of 'individualizing' as much as the activities of individuals consist in the daily reshaping and renegotiating of the network of mutual entanglements called 'society'....'individualization' consists of transforming human 'identity' from a 'given' into a 'task' and charging the actors with the responsibility for performing that task and for the consequences (also the side-effects) of their performance.[40]

This perennial activity of individualizing whilst navigating the complex, ever-shifting interdependences that constitute contemporary digital living, has reduced the social from a collective or mass to a "digital swarm," as expressed by Byung-Chul Han: "The digital swarm lacks the soul or spirit of the masses. Individuals who come

[40] Bauman, *Liquid Modernity*. Kindle Edition.

together as a swarm do not develop a *we.* No harmony prevails—which is what welds the crowd together into an active entity. Unlike the crowd, the swarm demonstrates no internal coherence."[41] This highlights the digital subject as simultaneously hyper-individualized yet hyper-networked, a position devoid of agency both individually and collectively.

This overview of contemporary digital subjectivity illustrates the complexities and shifts in fundamental constructs that are shaping individuals and society. These observations lend the foundational basis for assessing the implications of such conditions when applied to the digital immortal. Important questions regarding power structures, motivations, production and agencies, that have been highlighted in this chapter, acquire a new urgency when considering the dissolution and restructuring of relations between the key tenets of existence such as time, space, life and death, within the digital realm.

---

[41] Byung-Chul Han, and Erik Butler, *In the Swarm: Digital Prospects* (Cambridge, MA: MIT Press, 2017), 10.

Chapter Three

Implications of digital immortality

Digital immortality, when achieved through the memetic transduction of the bodily-bounded self to an un-bounded digital replicate, inevitably raises essential questions regarding the reproducibility of the self and the potential for an AI animated avatar to evolve beyond the parameters of what constitutes the individual self. This presents an inherent quandary for the endeavour of digital immortality. If the spatial, locational and experiential anchorage of the body is abolished, and the AI powered representational avatar is constantly optimized with exponential ability for learning, then, in what capacity is the individual preserved? And if the individual is not preserved, then what motive could digital immortality serve?

Martine Rothblatt's Terasem Movement, Inc. (a sister organization of The Terasem Foundation) has developed a project, titled "Cybernetic Beingness Revival" or "CyBeRev," dedicated to "test the comparability of single person human consciousness with a digital representation of the same person created by personality software that draws upon a database comprised of the original person's digitized interactions, as may be assessed by expert psychological review."[42] This site is similar in motivation to Lifenaut but functions as an experimental test ground. Rothblatt argues that with the development of computational consciousness, formational boundaries would maintain a human level cyber-consciousness, described as,

> a continuum of software-based human-level autonomy and empathy as determined by consensus of a small group of experts in matters of human consciousness [in the form of an advanced Turing test]... also including those kinds of independent thinking and feeling that occur subconsciously, in the unconscious mind. Cyberconsciousness software will have to provide for some quantum of unconscious conceptions, motivations, and decisions to produce a human-level mind…as code running in

[42] "WELCOME TO CYBEREV." *The Cyberev Project* (website), Accessed February 24, 2020. https://www.cyberev.org/.

> the background, of which a foreground information-processing unit is not "aware."[43]

If AI can be harnessed to function within the confines of a human mind through a coded limitation rather than the flesh of a brain, it is important to question the learning of this mind, if not through bodily experience. The supply of data from the original individual to the avatar mind would provide the foundation, but the "forever" of this form of immortality would experience and exist in the digital realm. This evokes questions of the environment and conditions that would be forming this human-computational mind, including the process of shaping the subconsciousness, as illustrated by Rothblatt. As mapped in the previous chapter, the digital realm is fraught with the obscuration of subjectivity and agencies, so if existence in the immortal form would constitute solely within this context, it is pivotal to assess the conditions and implications of the digital when progressively applied to an immortal entity. The interrelation between Silicon Valley, large tech companies and the immortality

[43] Martine Aliana Rothblatt, *Virtually Human: the Promise - and the Peril - of Digital Immortality* (New York: St Martins Press, 2014) Kindle edition.

endeavour, necessitates critical examination of the power structures that would govern this reality.

The organizations associated with the The Terasem Movement are all registered as not-for-profit charities with the purpose of educating the public, and do not solicit external funding except for the charitable donations of its users.[44] The foundation is funded by the founder, Martine Rothblatt, who is one of the highest earning CEOs in the biopharmaceutical industry, due to her position with the drug manufacturing company United Therapeutics.[45] The policy documents relating to Lifenaut and CyBeRev specify that the data collected from a user will not be sold or shared with any third parties. This, however, is not the case with the majority of platforms offering intentional immortality. The site eterni.me is a private, for-profit, company and states in its operational policy document that data will be shared with third party companies for the

---

[44] "What Is the Terasem Movement Foundation (TMF)?" *Terasem Movement Foundation* (website), Accessed February 24, 2020. https://www.terasemmovementfoundation.com/mission.

[45] Robert Hackett, "Highest-Paid Female CEO Seeks Immortality-Digitally." *Fortune* (September 12, 2014) Accessed February 24, 2020. https://fortune.com / 2014 / 09 /12/highest-paid-female-executive-seeks-immortality-digitally/.

purpose of analytics of user behavior, marketing and behavioral targeting.[46] These third party services utilized by eterni.me are predominately products of Google or Facebook, such as Google Analytics, Google Optimize, Google OAuth and Facebook Custom Audience. Eter9, also a for-profit private company, states in its privacy policy that it reciprocally shares data and analytics with third party companies such as YouTube, Wikiquote and Facebook.[47] It specifies, "this procurement [of data] is done by your counterpart. You may disable your counterpart activity in account settings, but in this case your ability to enjoy our services and interact with the network will be limited."[48] This highlights that immortality in the digital realm will still be embedded in the same data economy of "platform capitalism" that is governed by a monopoly of big tech companies such as Facebook, Inc and Alphabet, Inc. This also extends to the immortality endeavour offline, illustrating the degree of corporate interest in life extension technologies. This is

---

[46] *Eternime* (website) Accessed February 24, 2020. http://eterni.me/privacy.

[47] Help - Privacy. ETER9 (website), Accessed February 24, 2020. https://www.eter9.com/help/privacy.

[48] Ibid.

evidenced by Ray Kurzweil's employment as Director of Engineering at Google and the company's consequential establishment of the Calico biotech company, which intends to "solve death."[49] Calico is further fronted by Arthur D. Levinson, the current CEO of Apple, Inc. and former CEO of both Google and the large biotech company Genentech.[50] Further, the multinational corporation IBM, sponsors the Blue Brain Project that aims "to build biologically detailed, digital reconstructions and simulations of the rodent brain and, ultimately the human brain."[51] Another example of the big tech and immortality entanglement is Elon Musk's Neuralink corporation, dedicated to developing implantable brain–machine interfaces. These examples emphasise that any progression towards immortality, be it digital or material, will be thoroughly financialized with the central intention of capital gain. Thus, the

---

[49] John Naughton, "Why Silicon Valley Wants to Thwart the Grim Reaper." *The Guardian* (Guardian News and Media, April 9, 2017) Accessed February 24, 2020. https:// www.theguardian.com/commentisfree/2017/apr/09/ silicon-valley-wants-to-cheat-grim-reaper-google.

[50] "Calico." *Calico* (website), Accessed February 24, 2020. https://www. calicolabs.com/arthur-levinson/

[51] "Blue Brain Project." *EPFL* (website), Accessed February 24, 2020. https://www.epfl.ch/research/domains/bluebrain/.

transhumanist future will continue to be administered within a corporatocratic regime.

As detailed in the previous chapter, digital behaviors being data generative within the digital economy of data value and commodity, has ensured the alienation of the digital prosumer from their unintentional and immaterial abstract labor. As further highlighted in Chapter One, even the data bodies of the digital ghosts on Facebook continue to be mined and extracted for their profitable data. In the context of the digital immortal that functions as a continuing and autonomous data produced, and productive, being, this enacts the ultimate alienation, in a definitive fulfilment of the Marxian "commodity fetishism."[52] It is the literalization of the commodity taking on human properties, and human properties becoming the commodity, in a total subsumption of the capitalist paradigm.

The agency of the production of the commodity, which is the self, is alienated from the individual through the autonomous and artificially intelligent possession of

---

[52] Karl Marx and Friedrich Engels, *Capital: a Critique of Political Economy: The Process of Capitalist Production* (New York: Cosmo Classics, 2007), 81-96.

the immortal being. The optimization of AI and the feedback loop of data generated by an individual, which reproduces the environment and reflexively feeds back to the individual, ad infinitum, highlights the inevitable estrangement from the original source of the identity. The identity and environment of the immortal digital being would be completely dependent on the agenda and affiliations of the providers of the platforms enabling the immortalization. Jean Baudrillard writes in his work on *Symbolic Exchange and Death*:

> This is precisely the way in which power will later be instituted between the subject separated from its body, between the individual separated from its social body, between man separated from his labor: the agency of mediation and representation flourishes in this rupture…All the agencies of repression and control are installed in this divided space, in the suspense between a life and its proper end, that is, in the production of a literally fantastic and artificial temporality …All the future forms of alienation that Marx denounces, the separations and abstractions of political economy, take root in this separation of death.[53]

Immortality appears to be inherently bound to capitalist ideals particularly in its relation to accumulation and

[53] Jean Baudrillard, *Symbolic Exchange and Death* (London: Sage, 1976), 151.

reproduction, which is amplified in the digital realm. The endeavour to immortalize involves the accumulation of time and memory. The accumulation of time enables a colonization of the future by the ever-present, which is a vital condition of the immortal. This stockpiling of memory, or data, illustrates a privatized practice of accumulation, with emphasis on the individual's past instead of a collective history. In the unregulated time-space compression of the digital, this would result as an accumulation of aggregate moments, eternally reproducing a vibration between the past and the endlessly extending present. In the economy of the digital immortal as alienated data generator, this portrays a cyclical exercise of data production and reproduction in a perfect realization of self-generative surplus value: "Workers do not produce themselves: they produce a force independent of themselves. The success of this production, that is, the abundance it generates, is experienced by its producers only as an abundance of dispossession."[54]

[54] Guy Debord, *Society of the Spectacle* (New York: Zone Books, 1994) Np-section 31.

Although Rothblatt depicts a future in which humans and their digitally immortal counterparts live in hybridized unity, the reality that the digital entity will immortally occupy will remain in the virtual. Once the physical human dies, along with the interactions between the physical and digital version, the individual will theoretically transcend into a solely digital being, a transition that Kurzweil refers to as a "more capable" version of an individual, rather than an "old" and "new," reflecting the hierarchical neo-Darwinism of the transhumanist pursuit. The digital realm of infinite reproduction and pure representation that this entity would inhabit, aligns with Baudrillards' third order of simulacra, or "hyperrealism," where reduplication takes the place of cause and effect, or origin and end:

> The very definition of the real is *that of which it is possible to provide an equivalent reproduction.* It is a contemporary of science, which postulates that a process can be reproduced exactly within given conditions, with an industrial rationality which postulates a universal system of equivalences… At the end of this process of reproducibility, the real is not only that which can be reproduced, but *that which is always already reproduced:* the hyperreal.[55]

---

[55] Baudrillard, *Symbolic Exchange*, 94.

In the digital realm this "universal system of equivalences" is nullified, giving way to shifting parameters of infinite reduplication. In transhumanist rhetoric, this simulated reality would be a "utopian," free-market playground of on-demand switchable representations of bodies and environments, as written by Kurzweil:

> The Web will provide a panoply of virtual environments to explore. Some will be re-creations of real places; others will be fanciful environments that have no counterpart in the physical world. Some, indeed, would be impossible, perhaps because they violate the laws of physics. We will be able to visit these virtual places and have any kind of interaction with other real, as well as simulated, people (of course, ultimately there won't be a clear distinction between the two), ranging from business negotiations to sensual encounters…In virtual reality we won't be restricted to a single personality, since we will be able to change our appearance and effectively become other people. We can select different bodies at the same time for different people. So your parents may see you as one person, while your girlfriend will experience you as another. However, the other person may choose to override your selections, preferring to see you differently than the body you have chosen for yourself.[56]

This reality would be the total fulfilment of Baudrillard's

[56] Kurzweil, *The Singularity,* np.

hyper-reality, as an endlessly reduplicated existence devoid of cause and effect. The gamification of social world and reality would further comprise the posthuman rationale of “the end of subjectivity.” The Proteus effect, which observes the changing characteristics and behaviors of an individual in response to the appearance of their avatar in virtual worlds,[57] highlights that in a reality of limitless options for “embodiment,” postmodern fluidity would proceed as a dissemination and obliteration of identity. The nonlinearity of time and space comprising of aggregate moments would engulf the entity into only “being,” or existing, in those moments, without a trajectory of “becoming.” This dissemination and commodification of identity would essentially enact a total objectification of the immortal subject. The human perception of reality is integrally connected to our bodily senses. When the subject is untethered from the body, this instability of environment and embodiment would negate the preservation of the individual, thus nullifying the

---

[57] Jim Blascovich, *Infinite Reality: Avatars, Eternal Life, New Worlds, and the Dawn of the Virtual Revolution* (London: HarperCollins e-books) Kindle Edition, 102.

primary motivation for this form of immortalization.

Although these immortal avatars could not claim a history - as history requires the unfolding of time - they would still have a traceable digital trail. As detailed in the first chapter, the retention of information generated by our behaviors and usage patterns online constitute our digital existence. Being active in the digital realm as an operative node in the data economy equates to a visibility that enables enhanced surveillance and control. When this reality is reflexively constructed and governed in a late-capitalist corporatocracy, this control is all-pervasive. The author Shoshana Zuboff refers to this "rogue mutation of capitalism" as "surveillance capitalism:" "A new economic order that claims human experience as free raw material for hidden commercial practices of extraction, prediction, and sales; …[and] a parasitic economic logic in which the production of goods and services is subordinated to a new global architecture of behavioral modification."[58] The corporate omnipotence within the

---

[58] Shoshana Zuboff, *The Age of Surveillance Capitalism: The Fight for a Human Future at the New Frontier of Power (*New York: Public Affairs, 2019) Ebook.

virtual habitat of the digital immortal renders it nothing more than an ultra-visible commodity tool subjected to control and extraction in the absolute realization of this surveillance capitalist logic.

Ultimately, however, the notion of digital immortality appears a fallacy. The economic incentive of marketing the digital sphere as a utopian death-free realm reads as an insidious exploitation of humanity's greatest fear, for the comprehensive disclosure of information and data. The digital's dissolution of traditionally understood contra-distinctions extends to the notions of "mortality" and "immortality," as Bauman writes, "the struggle for eternal existence ends up as a rehearsal of non-being. Immortality, if ever it is to be achieved, shows all the abhorring features which made the transience and temporality repulsive and inspired the flight to the eternal. The dream of transcendence appears to be self-defeating."[59] This accounts for a digital entity that neither lives nor dies in a time-space compression, further eradicating "eternity." As evidenced at the beginning of

[59] Bauman, *Mortality*, 76.

this book, immortality has forever been a fantasy that structures the mediation of death in culture and society. Post-modern individualism, coupled with the late-stage capitalist migration to the digital realm, primed this essential death-evasion drive to be capitalized, resulting in a total tautological subsumption of the neoliberal capitalist paradigm.

Chapter Four

Recomposition of Being

Within the assumption that the pursuit of immortality and perceptions of death reflexively shape cultural and societal constructs, how can rethinking these relations impact both our digital and real world existence? The vital concerns of a digital immortality, as established in this book, are related to the diminished agency and identity of non-corporeality within the digitalized transference of economically motivated power structures. The drive for individually preserved immortality stems from capitalist individualism and the dogmas of accumulation and continuous growth. This is evidenced in the capitalization of death resulting in the multi-billion-dollar death and funeral industry, based on selling products and procedures for the unnecessary preservation of corpses, such as lead lined caskets and embalming.

However, the rising Death Positive movement is

challenging this industrial and capitalist dominance of death through de-stigmatizing the subject and reclaiming the narratives surrounding it. Through advocating natural burials, a central focus of the movement is to re-establish the importance of decomposition, emphasizing that through corporeal decomposition a universal immortality is achieved. The current industrialized and corporatized "traditional" death customs propagated by the funeral industry, such as preserving bodies and remains, is both an environmental and spatial problem. The withdrawal of bodies from the natural cycle of time and matter generates an accumulative stasis, as the continued stockpiling of preserved bodies colonizes both land space and time into an expanding present. This is relative to the existence of the digital body, which is not subject to the effects of time or space. The spatial and temporal delimitation of the digital provides an environment for infinite accumulation of digital bodies, and much like the preserved corpses in cemeteries, are beneficial only for generating capital for third party companies. Can a comparative overview of contemporary progressive death practices provide a model for rethinking digital interaction and behaviors regarding post-death digital existence, data generation and control? With a view to establishing the significance of decomposition as encompassing decay, regeneration and decentralization, I propose examining the impact of instituting time into the digital realm, and the potential reclamation of our data bodies.

When a body is naturally buried, it decays and recomposes into soil, which nourishes the earth and plant life. This natural cycle is inherently bound to an understanding of organic time. Digital decomposition is an already existing phenomenon that constitutes the corruption and decay of data. This fragility of the digital to the effects of entropic decay is often overlooked, particularly within the techno-optimist rhetoric of transhumanism, due to the contemporary anxiety of information loss corresponding to death in the digital realm. Digital decomposition occurs in two formats: mechanical and logical. Mechanical decay is related to the decomposition of the hardware through material erosion or the obsoletion of the hardware model. Logical decomposition is the decay of data through the degradation of the binary code. Such decomposition occurs due to updates in software and changes in how programming language operates. These updates shift the way in which information is read, corrupting code so it can no longer be processed. Digital information is dynamic, requiring constant maintenance in order to be preserved. Data decay is accelerated through massive accumulation of data that defies processing and protocols of prioritization. The speed at which information processing software is being developed leaves information vulnerable to the same obsoletion as its hardware counterpart. The financial repercussions of digital decomposition for companies dependent on big data is massive, and has therefore catalyzed the development of data cleansing systems,[60] with little doubt of a soon to be developed algorithm for optimized data preservation.

How can we consider the potential of data decomposition as a disruptive force to the capitalist

---

[60] 'The Four V's of Big Data.' *IMB Big Data & Analytics Hub* (website), Accessed April 6, 2020. www.ibmbigdatahub.com/infographic/four-vs-big-data.

hegemony over our data bodies, in order to instigate the notion of recomposition as an antagonist of accumulation? And would instituting an autonomously-controlled decomposition of data enable preservation of important data rather than the contemporary over-saturation which results in unregulated decay? The Death Positive re-institution of the cyclical planetary ecology of the organic matter of our bodies, when analogized to the digital, necessitates establishing an equivalent "common totality" of the digital realm. This would be comparable to the accessibility and expansion of knowledge, equally dependent on data. Would it be possible to propose an optional and periodical decomposition of an individual's surplus data that recomposes the beneficial components of that data anonymously into a larger body of knowledge? This would act as an inversion of the current emphasis on "analyzing people as data in order to deliver human consciousness up to data, rather than delivering data up to human consciousness,"[61] whilst further incentivizing

[61] Douglas Rushkoff, Ep. 128 Brewster Kahle "The Library of Everything," *Team Human.* (Podcast audio, April 24, 2019), Accessed April 6, 2020 https://teamhuman.fm/episodes/ep-128-brewster-kahle/

a common, contributive immortality that would lessen the anxiety of digital death and information loss.

Further, could collectivizing digital death towards recombinant decay provide a counteractive condition to the "romantic vitalism"[62]of techno-optimists and transhumanists that reflexively manifests economically as the ideology of free market self-regulation, or "spontaneous order?"[63] Extending decomposition into the digital sphere could potentially be considered a form of regulatory entropy, as an antithesis to cybernetic and algorithmic pattern predictability and capitalist expansion. This speculative proposition considers the notion of decomposition as relative to decentralization, which is imperative in fracturing the current corporate monopoly on information in the digital realm. It also works towards the establishment of a sustainable repository of knowledge. Here, re-composition can be regarded simultaneously as common redistribution.

---

[62] Peter Ludlow, *Crypto-Anarchy, Cyberstates and Pirate Utopias* (Cambridge: MIT Press, 2001), 356

[63] ibid.

Implementing the autonomous data retention of users and the abolition of corporate dominance in the digital realm would require a radical restructuring of digital infrastructure. To consider effective models of decentralization and redistribution of digital infrastructure relates to complexities regarding administrative governance and technological form, which extends beyond the purview of this book. However, thinking through the potential application of this proposition provides for provocative engagement with contemporary initiatives towards a more equitable digital existence. Decentralization of internet provision to a local level, instigated by community broadband initiatives, moves the foundational facility of the digital from a corporate monopoly, or duopoly, to public control. Transition to this strategy of municipally funded and locally maintained broadband, in itself, would dramatically alter the terrain of digital existence. Disabling corporate internet service providers would remove the financially incentivized prioritization of traffic towards certain content and services,[64] allowing for a bottom up decentralization and localization of these provisions.[65]

The community engagement aspect of these projects enhances the democratization of the digital through promoting literacy of digital realities and their innovative capacity, whilst further meeting local needs. It extends to greater control of a user's data through increased local agency for implementing regulation on data utility and procurement, through the cooperative model of service subscriber as service provider. Would it be possible for this local level digital provision to operate as a distributed node in a network that channels beneficial data into a centralized repository for an openly accessible archive of knowledge, functioning as a form of recomposition? Endeavours towards public broadband are motivated towards bridging the digital divide, through increasing connectivity of rural and lesser-resourced

---

[64] American Civil Liberties Union. *The Public Internet Option: How Local Governments Can Provide Network Neutrality, Privacy and Access for All* (report) Accessed April 6, 2020. https://muninetworks.org/ sites/ www. muninetworks.org / files / 2018-03-aclu_municipal_broadband_ report.pdf

[65] ibid.

communities. This would be fundamental in rethinking the larger body or knowledge as archive, through diversification and inclusivity of data, and its accessibility. Further, the cooperative framework of the community broadband model conforms to the proposed establishment of contributive cyclicality through its beneficial economic feedback into the community.[66] Implementing this model would instate a located temporal flow into the digital sphere as a parallel to the global connectivity. Would this alter the time-space compression of the contemporary digital, allowing for a situated containment of data bodies, thus greater agency? What would be the implications and potentials of such community based, bottom up internet on recombinant, collectivized digital immortality?

Introducing a semblance of cyclicality into the digital sphere has the potential to rupture the contemporary enterprise for accumulation of data and time. Emphasizing immortality as being achievable

---

[66] European Network for Rural Development. *New digital opportunities for rural areas: Community Broadband* (report) Accessed April 6, 2020. https:// enrd.ec.europa.eu / sites / enrd / files / tg_rural-businesses_case-study_ community-broadband.pdf

through collective maintenance and participation of this "common totality" would alleviate the current drive for individual reproduction and reduplication. Destigmatizing mortality and finitude through the reflexive alteration of societal concepts of time as linear would further complicate the notion of lineage. The transfer of significance toward an enduring universal project of knowledge sharing would facilitate the dismantling of accumulation through interrogating the practice of "inheritance." The concepts of individual decomposition and common recomposition provide the potential to shift post-death values from inheritance, differentiated as the linear, intergenerational amassing of wealth, to legacy, as a rhizomatic persistence of societal contribution.

This introduction of time into the digital sphere would further institute an unfolding of history, which is fundamental for societal learning and progression. This relates to the importance of grief, loss and resolution for human development, both as an individual and collective, as Simondon writes:

> Resolution of the psyche, though, cannot take place at the level of the individuated alone. It forms the basis of participation in a wider individuation, that of the collectivity. If the individual being puts itself, but nothing else, into question, then it will not be able to move beyond the limits of anxiety, for anxiety is a process without action, a permanent emotion that does not succeed in resolving affectivity, a challenge in which the individuated being explores the dimensions of its being without being able to progress beyond them.[67]

Fundamentally, death and loss imbue life with significance and are foundational to the human experience. The technological interrelation with human existence necessitates implementation of a form of organically influenced regulation within the immaterial sphere of the digital, in order to prevent digital technologies command of the complexities of humanity towards a reductive mechanistic essentialism. The initiation of a form of "death" in the digital, as instigating time, would instill a sense of significance to digital existence and, through its reflective capacity, potentially amplify our engagement with the material world through reinforcing our comprehension of bodily finitude.

[67] Gilbert Simondon. *The Genesis of the Individual*, Jonathan Crary & Sanford Kwinter (eds.), "Incorporations" (New York: Zone Books, 1992), 310.

Understanding immortality as greater than the self emancipates the individual from the solipsistic stagnation of space and being that burdens the quest for eternal life. Reclaiming death, both digitally and corporeally, reinstates subjectivity, as theorist Achille Mbembe writes

> The human being truly becomes a subject…in the struggle and the work through which he or she confronts death (understood as the violence of negativity). It is through this confrontation with death that he or she is cast into the incessant movement of history. Becoming subject therefore supposes upholding the work of death. To uphold the work of death is precisely how Hegel defines the life of the Spirit. The life of the Spirit, he says, is not that life which is frightened of death, and spares itself destruction, but that life which assumes death and lives with it.[68]

---

[68] Achille Mbembe and Steve Corcoran, *Necropolitics* (Johannesburg: Wits University Press, 2019), 14.

Conclusion

Death enacts a unification of the social, political and economic. Governing powers within the capitalist establishment have historically used death as a tool to economically colonize political and social spheres. These conditions are reified in the digital, underscored in the findings of this book, with the touting of digital immortality as simply a corporate exploitation of the primal death anxiety, for optimized extraction of valuable information and data. The sites for digital immortality are for-profit enterprises that produce capital in the data economy. They do this by sharing information with third party companies for speculative profiling of market needs. The unbound spatio-temporal hyper-reality of the digital enables a total subsumption of the capitalist paradigm into an abstracted and autopoiesistical system. The corporatocracy of the digital allows for large media conglomerates to monopolize the means and modes of production, as well as the product itself. This enacts the ultimate alienation as the worker becomes a digitally replicated autonomous data generator, and subjectivity itself has become the commodity.

Although the transhumanist drive towards immortality may seem genuine in its quest to solve death as the perceived universal source of pain and suffering, its immortality project is inherently bound to the capitalist political-economic framework. It is driven by the neoliberal doctrine of continuous expansion, accumulation and heroic transcendental individualism. The detachment of life from death, presented as oppositional conditions within a linear temporality, is a symptom of this transcendental rhetoric:

> The infinity of capital passes into the infinity of time, the eternity of a productive system no longer familiar with the reversibility of gift-exchange, but instead with the irreversibility of quantitative growth. The accumulation of time imposes the idea of progress, as the accumulation of science imposes the idea of truth: in each case, what is accumulated is no longer symbolically exchanged, but becomes an objective dimension. Ultimately, the total objectivity of time, like total accumulation, is the total impossibility of symbolic exchange, that is, death. Hence the absolute impasse of political economy, which intends to

> eliminate death through accumulation: the time of accumulation is the time of death itself. [69]

The immortality endeavor, as evidenced throughout this book, is a foundational supposition for the establishment of society. Conditioning this endeavor, is the prevailing ontological dominance of the ideology of "heroism." Heroism is indeed a crucial part of Western cultural hegemony, fundamentally rooted in the theistic traditions of heroic transcendentalism. Mythology heralds protagonists that overcome death in epics of katabasis, their immortality and heroism interchangeable. [70] This Western tradition of transcendentalism declares that everyone can be the immortal hero through devout behavioral conformation to ideological and belief systems that are mutually reinforcing; the monotheistic religions assurance of immortality in heaven, Capitalism's imperialist promise of accumulating wealth and, finally, the trust in the capacity of technology to ensure infinite life expansion. Becker writes, "the hope and belief is that the things that man creates in society are of lasting worth

---

[69] Baudrillard, *Symbolic Exchange*, 167.

[70] Most notable in the resurrection story of Jesus in the Christian tradition MISSING REFERNCE

and meaning, that they outlive or outshine death and decay, that man and his products count…."civilized" society is a hopeful belief and protest that science, money and goods make man count for more than any other animal. In this sense everything that man does is religious and heroic, and yet in danger of being fictitious and fallible."[71]

The congruence of these intertwined belief systems within the imaginaries of Western modernity has led to the societal emphasis on individualism, production and resolutional "progress," as the hero's prerogative is to always overcome and conquer. In the *Carrier Bag Theory of Fiction* (1986), the science fiction writer Ursula K. LeGuin contests these societally constructive storytellings of the "bashing, sticking, thrusting, killing" Hero protagonist, with a narrative that embodies the container, she writes:

> Conflict, competition, stress, struggle, etc., within the narrative conceived as a carrier bag/belly/box/house/medicine bundle, may be seen as necessary elements of a whole which itself cannot be characterized either as conflict or as harmony, since

[71] Becker. *Denial of Death,* 5.

> it's purpose is neither resolution nor stasis but continuing process. Finally, it's clear that the hero does not look well in this bag. He needs a stage or a pedestal or a pinnacle. You put him in a bag and he looks like a rabbit, like a potato.[72]

Le Guin's rendering of the anti-teleological anti-heroism is aligned with this book's proposition of reinstating a cyclicality as it relates to immanence, and as antithetical to transcendentalism.

Transcendentalism (of death) propagates a temporal linearity that endorses the logic of extractivist capitalism's disregard for life, death, others and the complex forms of worlds and worldings we create and inhabit. Immanence, on the other hand, is defined as, simply put, 'the condition of being entirely within something.'[73] This infers a non-hierarchal interconnectivity. It is the positioning of the individual in a rhizomatic network of being that simultaneously relieves the alienating and destructive pursuit of heroism, whilst instituting embedment in a communal responsibility of care. Immanence establishes a temporal cycle that situates a duty to both the past and the

---

[72] Ursula. K. Le Guin, *Dancing at the Edge of the World: Thoughts on Words, Women, Places* (New York: Grove Press, 2006), 169.

[73] MISSING SOURCE.

future as inherent to the present, and life and death as indissociable in the continuum of being, or, as suggested by LeGuin, a continuous process. How would this kind of shift in societal ontological conception of continuity impact, or perhaps utterly negate the need for diverse forms of memorialization as well as pursuits for immortality? And what would be the potential of freeing technology from its epiphylogenetic function, as proposed by Stiegler? I propose that in a time of the hyper-real and continuous abstract reproduction, death could be seen as a unifying anchor to immanent materiality.

The quest for extension of life and immortality has historically been the pursuit of the wealthy, as the reflexive endeavor for continuous expansion and accumulation. From the memetic immortality aspirations of commissioning extravagant monuments and portraits, initiating epic wars and imperial conquests, to the contemporary life enhancement efforts of extensive supplement regimes and intravenous longevity treatments,[74]

---

[74] See rejuvenation attempts through young blood transfusions. Danny Fortson, "Blood extract of teenager on sale at just £6000 a shot." *The Times* (August 20, 2017) Accessed April 6, 2020. https://www.thetimes.co.uk/

digital replication and brain uploading technologies,[75] or cryogenic suspension and medical nanorobotics.[76] The capitalist amassing of wealth as a means to fund more elaborate and expansive situations of heroic protaganism is a premise for these endeavors to colonize, engulf, and extract from, the lives of others. As Le Guin writes:

> Heroes are powerful. Before you know it, the men and women in the wild-oat patch and their kids and the skills of the makers and thoughts of the thoughtful and the songs of the singers are all part of it, have all been pressed into service in the tale of the Hero. But it isn't their story. It's his.[77]

In the current digital corporatocracy that pervades life both online and offline in an accelerated capitalist paradigm, this all pervasive extractivist colonization of the lives of

---

edition / news / blood-extract-of-teenager-on-sale-at-just-6-000-a-shot-gz3t8pqrw; Jocelyn Kaiser, "Young blood antiaging trial raises questions." *ScienceMag*, Aug 1, 2016. Accessed April 6, 2020. https:// www.sciencemag.org / news / 2016 / 08 / young-blood-antiaging-trial-raises-questions

[75] See Nectome startup, $10,000 deposit just to join the waitinglist: *Nectome* (website), Accessed April 6, 2020. https://nectome.com/about-us/; Antonia Regalado, "Nectome will preserve your brain, but you have to be euthansied first." *MIT Technology review* (March 13, 2018), Accessed April 6, 2020. https://www.technologyreview.com/2018/03/ 13 / 144721/a-startup-is-pitching-a-mind-uploading-service-that-is-100-percent-fatal/

[76] See Aubrey de Grey's Alcor Life Extension Foundation offering cryogenic suspension: *Alcor Life Extension Foundation* (website), Accessed April 6, 2020. https://alcor.org/BecomeMember/scheduleA.html

[77] Le Guin, *Dancing on the Edge*, 166.

others is absolute, as capitalism fulfils its teleological totality as the ultimate hero, narrative, and storyteller.

Acknowledgements

This book may not be very long, but it's feels like it's been a long time in the making, with invaluable support from so many. First, this dissertation was only possible with thanks to the indispensable guidance and mentorship of Antonia Majaca, who's teaching and energy reinvigorated my relationship to theory in a way that made the journey with this research so much more of a pleasure. Samantha McCulloch for her editing prowess and truly infinite seeming support, counsel and critical conversation. To Felix Bamforth for the beautiful cover design. My thanks always to the encouragement and input from Edel O'Reilly, whose ability to synopsize everything; from this dissertation to all of life's most complex phenomena, is a constant source of learning and laughter. Polly Wright, for being a constant cheerleader, and whose curatorial practice of sensitively mediating complex subjects through creative and collaborative thinking has been an inspiration. With deep appreciation to the various contributions, conversations and encouragements from Clementine Edwards, Mariusz Bogacki, Eric Peter, Marina Vishmidt, Bassam El Baroni, Lukasz Hoffman and Jasmin Schaedler. The Dutch Art Institute for being a critical yet supportive academic environment to which I owe the ongoing conversations and friendships that have become gladly inextricable from all facets of my work and life. Most importantly, my eternal gratitude to my parents, that through both their lives and deaths continue to guide me though a vast spectrum of emotion and existential reckoning that informs all my research and being in the world. And finally, but equally as importantly, with loving thanks to both my brothers, David and Milo, for just always keeping it real.

Bibliography

Baudrillard, Jean. *Symbolic Exchange and Death.* Los Angeles: Sage Publishing, 2017. Print.

Bauman, Zygmunt. *Mortality, Immortality and Other Life Strategies.* Cambridge: Polity Press, 1992. Print.

Bauman, Zygmunt. *Liquid Times.* Cambridge: Polity Press, 2007. Kindle Edition.

Bauman, Zygmunt. *Liquid Modernity.* Cambridge: Polity Press, 2012. Kindle Edition.

Becker, Ernest. *The Denial of Death.* New York: The Free Press, 1973. Print.

Bernard, Andreas, and Valentine A. Pakis. *The Triumph of Profiling: the Self in Digital Culture.* Cambridge, UK: Polity Press, 2019. Ebook.

Blascovich, Jim. *Infinite Reality: Avatars, Eternal Life, New Worlds, and the Dawn of the Virtual Revolution.* London: HarperCollins e-books. Kindle Edition.

Clough, Patricia Ticineto. *The User Unconscious: On Affect, Media, and Measure.* Minneapolis: University of Minnesota Press, 2018. Kindle Edition.

Debord, Guy. *Society of the Spectacle.* New York: Zone Books, 1994. Print.

Elias, Norbert, and E. F. N. Jephcott. *The Civilizing Process: Sociogenetic and Psychogenetic Investigations.* New York: Urizen Books, 1977. Print.

Han, Byung-Chul, and Erik Butler. *In the Swarm: Digital Prospects*. Cambridge, MA: MIT Press, 2017. Ebook.

Haraway, Donna Jeanne. *Cyborg Manifesto*. Victoria, British Columbia: Camas Books, 2018. Print.

Harvey, David. *The Condition of Postmodernity*. Cambridge, MA: Blackwell, 2000. Print.

Hayles, Nancy Katherine. *How We Became Posthuman: Virtual Bodies in Cybernetics, Literature and Informatics*. Chicago: Univ. of Chicago Press, 2010. Print.

Hviid Jacobsen, Michael ed. Postmortal Society: Towards a Sociology of Immortality (Studies in Death, Materiality and the Origin of Time). London and New York: Taylor and Francis, 2017. Kindle Edition.

Kurzweil, Ray. *The Singularity is Near: When humans transcend biology*. New York: Viking, 2005. Print.

Kurzweil, Ray. *How to Create a Mind*. London: Duckworth Overlook, 2013. Kindle Edition.

K., Le Guin Ursula. *Dancing at the Edge of the World: Thoughts on Words, Women, Places*. New York: Grove Press, 2006.

Marx, Karl, and Friedrich Engels.. *Capital: a Critique of Political Economy: The Process of Capitalist Production*. New York: Cosmo Classics, 2007. Print.

Mbembe, Achille, and Steve Corcoran. *Necropolitics*. Johannesburg: Wits University Press, 2019. Print.

Rothblatt PhD, Martine. Virtually Human. New York: St. Martin's Press, 2014. Kindle Edition.

Russell, Stuart. *Human Compatible: Artificial Intelligence and the Problem of Control.* London: Penguin Books, 2019. Print.

Simondon, Gilbert. *The Genesis of the Individual,* Jonathan Crary & Sanford Kwinter (eds.), "Incorporations" (New York: Zone

Books, 1992): 297–319. Print.

Stiegler, Bernard. *Technics and Time. Disorientation*. California: Stanford University Press, 2009. Print.

Zuboff, Shoshana. *The Age of Surveillance Capitalism: The Fight for a Human Future at the New Frontier of Power*. New York: Public Affairs, 2019. Ebook.

Index